NATURE'S Secret SUPERHEROES

BATS

THE DARK-WINGED CRUSADERS

Stephanie Feldstein

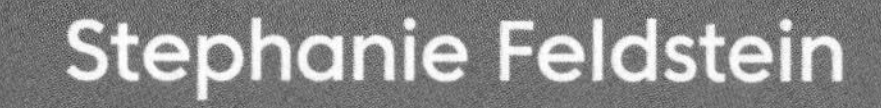

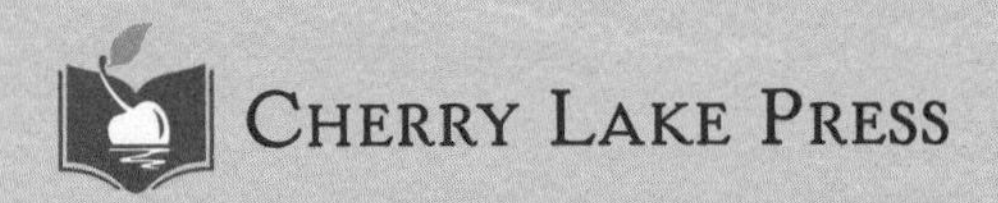

Published in the United States of America by Cherry Lake Publishing Group
Ann Arbor, Michigan
www.cherrylakepublishing.com

Reading Adviser: Beth Walker Gambro, MS, Ed., Reading Consultant, Yorkville, IL
Book Designer: Dan Winter

Photo Credits: © Rudmer Zwerver/Shutterstock, cover, title page; © SAHACHATZ/Shutterstock, 6; © Passakorn Umpornmaha/Shutterstock, 7; © Ienjoyeverytime/Shutterstock, 7; © Corina Daniela Obertas/Shutterstock, 9; © Dave Montreuil/Shutterstock, 11; © Griffin Gillespie/Shutterstock, 12; © Gulf MG/Shutterstock, 13; © Ienjoyeverytime/Shutterstock, 14; © Seregraff/Shutterstock, 15; © KAWYTHAIDOTCOM/Shutterstock, 17; © Danita Delimont/Shutterstock, 18; © Maris Grunskis/Shutterstock, 20; © Nitin Chandra/Shutterstock, 21; © Ienjoyeverytime/Shutterstock, 22; Rachel Harper, Bat Conservation International, 23; Photo courtesy Al Hicks, New York Department of Environmental Conservation, 25; Photo courtesy of Pete Pattavina, U.S. Fish and Wildlife Service, 26; © Ienjoyeverytime/Shutterstock, 28; © Lois GoBe/Shutterstock, 29; © Ienjoyeverytime/Shutterstock, 30; © Heather Wharram/Shutterstock, 31

Cherry Lake Press is an imprint of Cherry Lake Publishing Group.

Library of Congress Cataloging-in-Publication Data

Names: Feldstein, Stephanie, author.
Title: Bats : the dark-winged crusaders / written by Stephanie Feldstein.
Description: Ann Arbor, Michigan : Cherry Lake Publishing, [2025] | Series: Nature's (secret) superheroes | Audience: Grades 4-6 | Summary: "This title from our Nature's (Secret) Superheroes series delves into the powerful effect bats have on their environment. Learn how these underappreciated and often misunderstood environmental superheroes work to keep their ecosystems going strong"-- Provided by publisher.
Identifiers: LCCN 2024035978 | ISBN 9781668956366 (hardcover) | ISBN 9781668957219 (paperback) | ISBN 9781668958087 (ebook) | ISBN 9781668958957 (pdf)
Subjects: LCSH: Bats--Ecology--Juvenile literature. | Bats--Juvenile literature.
Classification: LCC QL737.C5 F435 2025 | DDC 599.4--dc23/eng/20240815
LC record available at https://lccn.loc.gov/2024035978

Cherry Lake Publishing Group would like to acknowledge the work of the Partnership for 21st Century Learning, a Network of Battelle for Kids. Please visit Battelle for Kids online for more information.

Printed in the United States of America

Note from publisher: Websites change regularly, and their future contents are outside of our control. Supervise children when conducting any recommended online searches for extended learning opportunities.

TABLE OF
CONTENTS

INTRODUCTION

ADVENTURES IN CAVES

Bracken Cave is near San Antonio, Texas. The land around the cave used to be grazed by cows. Now it's a nature preserve to protect bats. It's home to the biggest bat **colony** in the world. As many as 20 million bats live in the cave during the summer.

Bracken Cave is used by Mexican free-tailed bats. Pregnant bats **roost** there together. They give birth and care for their pups in the cave.

Mexican free-tailed bats eat flying insects like flies and beetles. A big colony of Mexican free-tailed bats can eat 250 tons of insects a night. That's more than the Statue of Liberty weighs! The bats' favorite food is moths. They eat a lot of corn earworm moths. These moths attack crops like artichokes and watermelons. By eating so many moths, the bats help save the crops from being ruined.

Bats all over the world stop insects from taking over crops and forests. Some bats help plants grow. They're an important part of **ecosystems** on every continent except Antarctica.

Up to 20 million Mexican free-tailed bats live in Bracken Cave. Bats fly out of the cave every evening.

Ecosystems are places where plants, animals, and the environment rely on each other. They help produce food, shelter, water, and clean air. Every kind of **native species** is important. If one disappears, it can harm the whole ecosystem. But some species are secret superheroes. They play an especially big part in shaping the web of life.

It's not easy being a superhero. Superhero species face a lot of danger. Some are even at risk of **extinction**. Extinction is when all of one kind of plant or animal dies. Bats remind people of Halloween and vampires. A lot of people misunderstand bats. Sometimes bats are killed because people are afraid of them.

But a world without bats is much scarier. We can all help save the day so nature's superheroes can keep protecting the planet.

BAT TALES

Bruce Wayne was afraid of bats. But when he became a superhero, he turned his fear into a strength. He became Batman.

During the day, Bruce Wayne is known for being the richest man in Gotham City. At night, he becomes Batman. Real bats are also nocturnal. That means they rest during the day and are active at night.

Bats' wings catch the air and help them glide above it.

Batman doesn't have superpowers. But he's known for being very smart and having a good memory. Bats also have good memories. They remember where to find the best food. Scientists found that bats can remember sounds for years.

Like many kinds of bats, Batman uses a cave. The Batcave is full of gadgets that help Batman fight crime. Batman uses night vision lenses and other technology to help him get around in the dark like a bat.

Batman uses his wing-shaped cape to glide through the air. But he can't fly like a real bat. Other mammals like flying squirrels also glide short distances. But bats are the only mammals that can really fly. They flap their wings to power their flight.

CHAPTER 1

ORIGIN STORY

More than 1,400 different kinds of bats live on Earth. In fact, one out of five species of the world's mammals are bats. Most bats are small enough to fit in your hand. The smallest bat is the bumblebee bat. It's smaller than an adult's thumb. The biggest bat in the world is the giant golden-crowned flying fox. Its wingspan is more than 5 feet (1.5 meters) long. Some bats can fly faster than a cheetah runs. Cheetahs can sprint as fast as 75 miles (121 kilometers) per hour. But Mexican free-tailed bats can fly almost 100 miles (161 km) per hour.

Bats live almost everywhere in the world. They're in forests and cities. They're in deserts and tropical rainforests. Not all bats live in caves. Some live in trees. Others live in spaces between rocks. They also live in houses or under bridges. They roost in cool, dark places during the day. They hunt all night until it's almost dawn.

The bumblebee bat is also known as the Kitti's hog-nosed bat. It weighs less than a penny!

Almost all bats hang upside down. Bats' feet clench like fists when they're relaxed. That's how they sleep on the ceiling of caves. When they want to fly, they let go. They fall to pick up speed. Then they fly into the night.

Different bats eat different kinds of food. All bats in the United States are **insectivores**. They eat mostly insects. Some bats in other parts of the world drink nectar from flowers. Others eat fruit. Some bats have specific favorite foods like fish or frogs.

Bats usually drop about 2 to 3 feet (0.6 to 0.9 m) to pick up enough speed to fly. It depends on their size. ▶

DID YOU KNOW?

Echolocation is one of bats' superpowers. A bat makes sounds through its nose or mouth. The sound is so high-pitched that human ears can't hear it. The bats listen to how it echoes back. They can tell the difference between an insect and the leaf it's sitting on based only on the echo. They can tell where tiny insects are flying. Half a second later, they swoop in and catch them. They can hear the footsteps of a beetle from 6 feet (1.8 m) away.

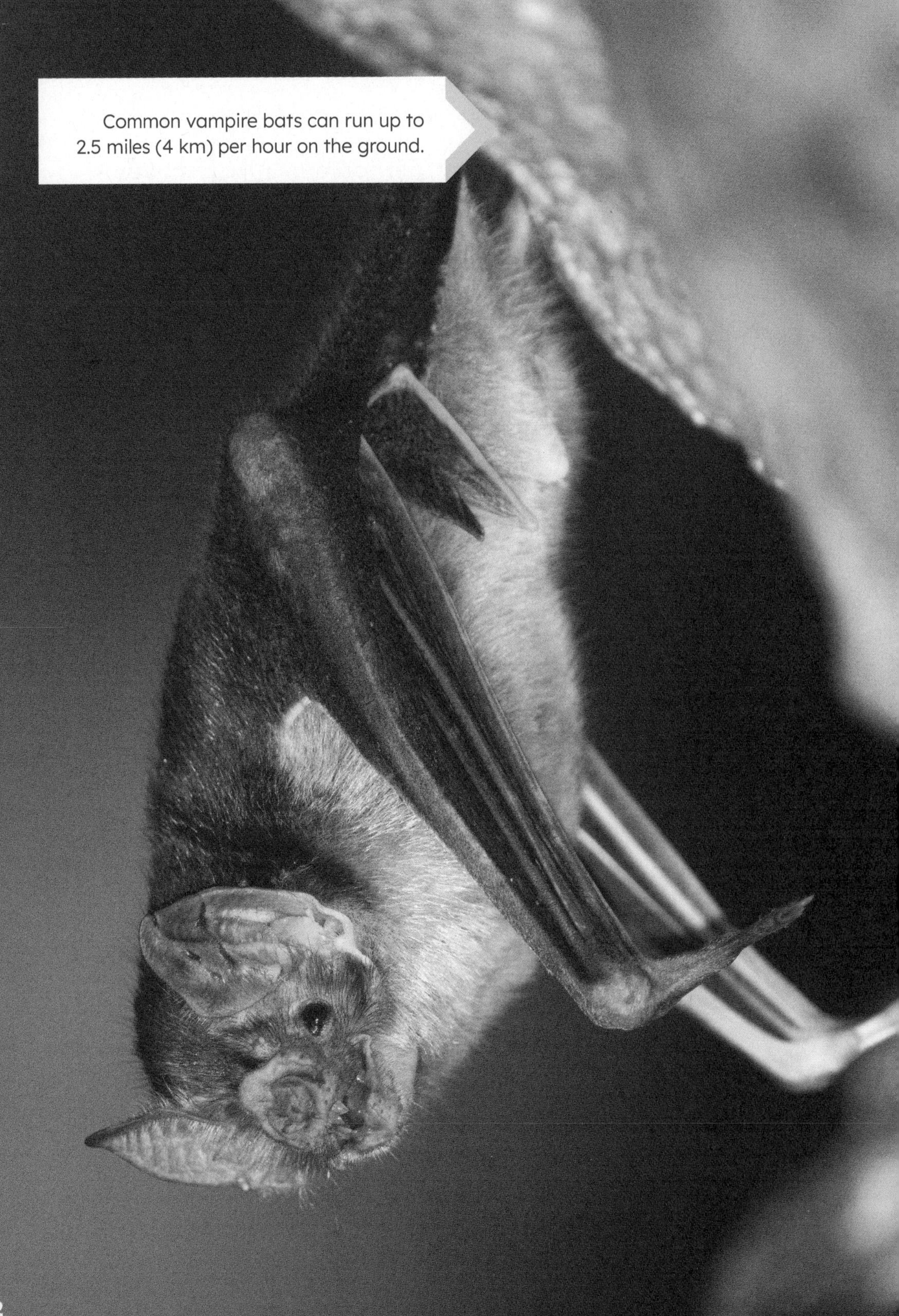

Common vampire bats can run up to 2.5 miles (4 km) per hour on the ground.

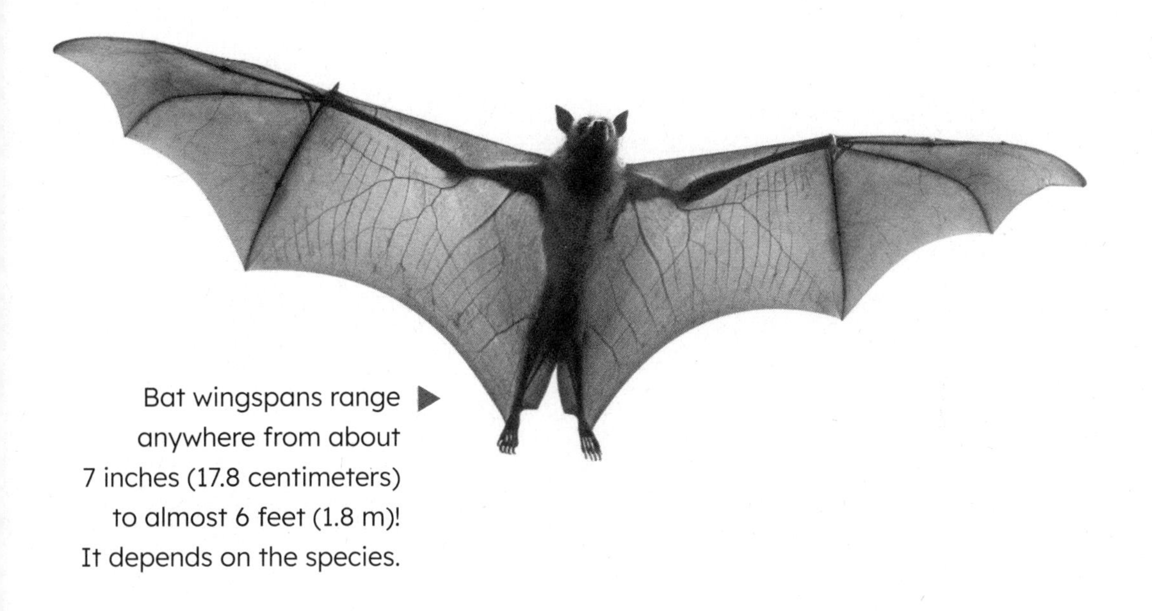

Bat wingspans range anywhere from about 7 inches (17.8 centimeters) to almost 6 feet (1.8 m)! It depends on the species.

There are only three kinds of vampire bats. They live in Latin America. Only one kind of vampire bat feeds on the blood of mammals. They usually eat from cattle. The other two kinds feed on the blood of birds. None of them turn anyone into vampires.

The saying "blind as a bat" isn't true. Bats may have small eyes, but some of them see three times better than humans. Some bats may even be able to see colors at night.

FAMILY TREE

Scientists don't agree on how to group all the different kinds of bats. They used to split them into megabats and microbats. Megabats were bigger bats that ate fruit. Microbats were smaller and used echolocation. Echolocation is when animals use sound to figure out what's around them. It helps bats find their way in the dark. It also helps them find food.

But scientists discovered that some bats that are closely related to megabats are small. They found that the ancestor of all bats might have used echolocation. Different scientists now group bats in different ways.

It gets even more confusing when you go farther up the family tree. Scientists learn how animals are related by looking at fossils. They look at animals living today. They compare how their bodies work and how they behave. They piece together all of that information like a puzzle. That's how they figure out which animals had a common ancestor hundreds of millions of years ago.

Even though they look a bit like rats, bats live much longer. A bat can live up to 35 years! Rats usually only live up to 2 years.

People often think bats are rodents because they look like mice with wings. But they're more closely related to humans than rodents. Because humans and bats share a family tree, bats could hold secrets that can help people. Scientists study bats for ways to make human medical treatments better.

CHAPTER 2

HOW BATS SAVE THE DAY

One bat can eat thousands of insects in a single night. They can eat up to half their body weight in bugs. If they're pregnant or feeding a newborn pup, they can eat their whole body weight in bugs!

Without bats, there would be a lot more insects. But more mosquito bites wouldn't be the only problem caused by bats' absence. Without bats, insects could take over ecosystems. It would cost farmers a lot of money in lost crops. Bats provide pest-killing services worth more than $23 billion a year.

Bats' appetites also help people use fewer **pesticides**. Pesticides are toxic chemicals made to kill a plant or animal that might harm crops. But they're so poisonous that they kill other wildlife, too.

Bats don't fly in formations like birds do. They usually fly in a straight line, unless bugs or other things get in the way!

DID YOU KNOW?

Echolocation isn't just a superpower for land. Dolphins and whales use echolocation underwater. The sounds they send out are so loud that their ears are shielded to protect them from the echo. Dolphins can sense something as small as a golf ball almost a football field away.

Pesticides kill bees and butterflies. They kill birds. They wash into streams and hurt fish and frogs. Bats help lots of different kinds of wildlife by eating so many bugs.

Bats also help plants. Some bats are **pollinators**. Pollinators are animals that move pollen between plants. That helps flowering plants produce seeds to grow new plants.

Other bats eat fruit. When they poop out the seeds, it helps rainforests grow. Bats poop while they're flying. It helps scatter the seeds across clearings. Bats help as much as 95 percent of new forests grow after trees have been chopped down.

◀ Most pollinator bats live in desert and tropical climates. Lesser long-nosed bats and Mexican long-tongued bats are two species that can be found in the United States.

Bat poop is called **guano**. Some people use guano as fertilizer to help plants grow. But removing guano from caves can harm the ecosystem. Animals like salamanders and crayfish often live in caves with bats. These animals feed on guano.

Bats help keep people healthy, too. They eat bugs that can spread human diseases. Vampire bats have special saliva for feeding on blood. Their saliva has helped develop medicines to prevent strokes. Vampire bats have helped save human lives.

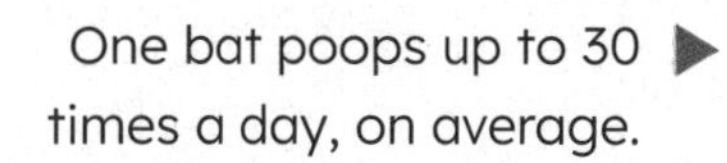

One bat poops up to 30 times a day, on average.

Bats eat all kinds of bugs. Some of their favorites? Moths, mosquitoes, beetles, crickets, grasshoppers, and stink bugs.

SUPER SIDEKICK

DR. MELQUISEDEC GAMBA-RIOS

Melquisedec Gamba-Rios grew up in Colombia and Costa Rica. He loved exploring the outdoors. As a kid, he didn't think he'd grow up to be a scientist. He didn't know scientists could work outside.

Gamba-Rios started studying biology in college. One of his first projects was helping a friend release bats into nature. He was still learning how to handle bats. One of them bit him. He learned that he should have been wearing gloves. But he also learned that bats were fascinating.

Today, Dr. Gamba-Rios works to protect bats in Miami. He studies Florida bonneted bats. They're the rarest bats in the United States. They get their name from their big ears that look like bonnets. They live only in a few places in South Florida. They're rapidly losing habitat to growing cities. Hurricanes are destroying the places where they roost. They're in danger of extinction. Dr. Gamba-Rios is working to save them.

Dr. Gamba-Rios works with an organization called Bat Conservation International. They do a lot of good work for bats!

Dr. Gamba-Rios collects data to better understand Florida bonneted bats. He looks at where they live in cities. He looks at how they find food. Then he works with others to create a plan to protect the bats.

CHAPTER 3

THE BAT'S NEMESES

Bats face a lot of the same threats as other wildlife. They're losing habitat to growing cities and farms. Their ecosystems are harmed by climate change. But one of the biggest threats to North American bats is found in their caves.

More than half of North American bats hibernate. **Hibernation** is when animals save energy to survive cold weather. They slow their breathing. They even slow their heart rates. They can go for long periods without food. Bats tend to hibernate in caves because the temperature stays steady through the winter.

In 2006, cave explorers in New York started to notice bats with white fuzz on their noses. Soon scientists started to see sick and dying bats with white noses. They realized it was a fungus.

A little brown bat in New York has white fungus on its nose.

DID YOU KNOW?

More than 200 kinds of bats around the world are in danger of extinction. In North America alone, more than half of the bat species could be in trouble. But those are just the bats scientists have been able to study. More than 230 bat species exist that scientists know little about. Scientists need more information about these species to know if they're also in trouble.

A tricolored bat in Cloudland Canyon State Park, Georgia, shows symptoms of white-nose syndrome.

A fungus is a living thing. It's not a plant or an animal. Mushrooms and mold are kinds of fungus. The fungus making bats sick grows in cool, damp places. That's exactly where bats hibernate in big groups.

The fungus causes a disease called white-nose syndrome in bats. It's very deadly. It kills between 70 and 90 percent of bats hibernating together. In some cases, it's killed entire colonies.

Bats can pick up the fungus from the caves where they roost. They can pass it between each other. It can also travel on people's clothes. The fungus is in caves across the United States now. Millions of bats have died. It's one of the worst wildlife disease outbreaks in North American history.

There's no cure for white-nose syndrome. The best way to save bats is to stop the spread. The fungus can stay on clothes for years, even if they've been washed. Keeping people away from places where bats hibernate can help. Many caves and abandoned mines have been closed. People who go into caves are warned not to wear those same clothes in any other caves, even if the clothes have been washed.

HOW TO HELP SAVE THE DAY

Bats are misunderstood. Some people think they're scary. Some people think all bats are vampires. Fear puts bats in danger. If people are afraid of bats, they may try to harm them. They won't help protect them. You can help by teaching other people about bats. Talk about how bats are secret superheroes.

You can do many things to protect bats around your house. Lights can stop bats from flying around and eating. You can turn off outdoor lights at night. It's especially important at dusk when the most insects are out.

Don't use pesticides around your house that might harm bats or the insects they eat. Go one step further and plant a pollinator garden. A pollinator garden uses native plants. It attracts helpful insects. It attracts bugs that bats like to eat.

A pollinator garden is just a garden that includes flowers with nectar and pollen. Native plants will be familiar and yummy for bugs.

Some people think bats carry diseases. But lots of animals carry diseases that can make people sick. As long as bats and their habitat are left alone, people aren't likely to catch anything from them. If you find a bat in your house or on the ground, don't touch it. Call your local wildlife rescue to safely remove it. Bat World Sanctuary has a map of bat rescuers on their website. They also have tips on what to do if you find a bat.

ACTIVITY

WELCOME BATS TO YOUR NEIGHBORHOOD

Bats live across the United States. Many bats spend the summer roosting in trees, old buildings, or under bridges. There may not be many spots near you that make the perfect bat home. But you can hang a bat house to help them out.

Bat houses are wooden boxes. They have narrow rooms inside. That makes it cozy like the space between bark and a tree trunk where some bats like to sleep. The houses are painted dark inside to keep pups warm. The surfaces are rough so bats can climb inside them.

1 BUILD A BAT HOUSE

Bat house kits come with precut wood that you put together and paint. Search online with an adult for bat house kits certified by Bat Conservation International. That way, you can be sure the design is exactly what bats need. You can find certified bat houses that are already built. You can also find plans to build a bat house from scratch.

2 HANG YOUR BAT HOUSE

It's important to hang your bat house in the right spot. You don't want to hang it on a tree. That makes it too easy for other animals to get to it. There also needs to be 15 to 20 feet (4.6 to 6.1 m) of clear area below the house. That's the bats' runway. They need space to drop out of their house. Dropping helps them pick up speed to take flight.

The best place to put a bat house is on the side of a human house. Your house helps keep their house warm. The roof helps protect them. You can also put a bat house high up on a pole.

3 WATCH THE BATS

It may take bats a little time to find their new house. Keep an eye on it around nightfall. You're most likely to see bats in the summer when it's just getting dark. That's when there are the most insects for them to catch.

LEARN MORE

BOOKS

Carson, Mary Kay. *The Bat Scientists.* New York, NY: Clarion Books, 2013.

Feldstein, Stephanie. *Save Pollinators.* Ann Arbor, MI: Cherry Lake Publishing, 2023.

Gish, Melissa. *Bats.* North Mankato, MN: The Creative Company, 2024.

Gray, Susan H. *Gray Bat.* Ann Arbor, MI: Cherry Lake Publishing, 2008.

Gregory, Josh. *From Bats to . . . Radar.* Ann Arbor, MI: Cherry Lake Publishing, 2013.

Markle, Sandra. *The Case of the Vanishing Little Brown Bats.* Minneapolis, MN: Millbrook Press, 2015.

GLOSSARY

colony (KAH-luh-nee) a group of bats living together

echolocation (eh-koh-loh-KAY-shuhn) use of sound to figure out the location of distant or invisible objects

ecosystem (EE-koh-sih-stuhm) a place where plants, animals, and the environment rely on each other

extinction (ik-STINK-shuhn) when all of one kind of plant or animal dies

guano (GWAH-noh) bat poop

hibernation (hye-buhr-NAY-shuhn) becoming inactive to save energy and survive cold weather

insectivore (in-SEK-tuh-vor) an animal that eats mostly insects

mammals (MAA-muhlz) animals that have a backbone, grow hair or fur, breathe air, and produce milk for their young

native species (NAY-tiv SPEE-sheez) plants and animals that are a natural part of an ecosystem

nocturnal (nahk-TUHR-nuhl) resting during the day and being active at night

pesticides (PEH-stuh-sydz) toxic chemicals made to kill plants or animals that might harm crops

pollinators (PAH-luh-nay-tuhrz) animals and insects that move pollen between plants

roost (ROOST) to settle down for sleep or rest

INDEX